HVAC GUIDE BOOK

Practical manual for repairs and maintains of HVAC installations

Aria G. Mason

Table of Contents

CHAPTER ONE

INTRODUCTION FOR HVAC

HVAC is time duration stemmed from Heating, Ventilating, and Air Conditioning, and it is a desktop that is aimed to provide comfort for indoor areas.

Science contributes to identification of vary of measures for that treatment and thermal control, and HVAC buildings will be engineered notably primarily based on the favored design points. Cooling buildings fluctuate from large air conditioning constructions to small portable ones that furnish air cooling. These constructions come in use for heat climates or spaces. What these buildings do is that they leave out the air over fluid-cooled surfaces before than fundamental it into the supposed place for cooling.

How HVAC draw heats to convert it to cool air (recycling)

The air is exchanges warmness with the cooled flooring and then is directed to the vicinity turning in its duty. The cooling fluid ought to be water or an evaporative fluid. In case of evaporative systems, due to the reality the manner of evaporation is endothermic, i.e. it absorbs heat, and a lot of heat can be absorbed with the resource of the evaporator providing immoderate tiers of cooling. Of course, such fluids cannot be built-in to all buildings due to financial, safety and environmental concerns.

Thermal radiations device in heating circle

Heating can be carried out increased barring situation and there are more techniques for establishing a heating system. Heating can be completed with the aid of the usage of electric powered heating, heat pumps that ought to moreover be used for cooling increased on this in each and every different post, water, steam, and heat fuel radiators, etc. Heating constructions ought to provide heat air via the usage of pushing it surpassed some heat flooring into the supposed vicinity for heating, or ought to be done by using thermal radiation from heat surfaces.

Tubes and floor connection underneath

Surroundings pleasant way for heating vicinity would be floor heating through radiation from water or glycol carrying tubes hooked up underneath the flooring of a floor. Since the hotter air would waft above the cooler air and warmness grant is on the floor, this method would be very efficient.

How to service HVAC

Inspect ducts, vents, cabinet, refrigerant lines, and the perimeter of the unit for debris, mold, leaks, charge.

Clean all parts

Cleaning the condenser and evaporator, cleansing filters or

alter them if necessary. Cleaning drain traces and pans, checking humidity tiers as precise as thermostat and controls for relevant measurement and regulations, Inspecting fan motor, blowers, and blades.

Oil parts some component

Lubricating moving parts, Replacing worn-out belts and pulleys, Inspecting electrical elements and connections as excellent as altering batteries,

CHAPTER TWO

INSTALLATIONS

Installations Of Electric Heat Radiant

Have a measurement

Carefully measure your room, generally to the walls. Use diagram paper to draw your floor graph with your measurements. Wire-embedded mats are slim and can be minimize to inform round obstructions in small rooms. When measuring, make sure you precisely embody the neighborhood of vanities, floor vents, tubs and relaxation room drains in your drawing.

Wire connect needed device

Heat mats or wires want to no longer be positioned underneath eternal fixtures or zero-clearance furniture. Include in the drawing the placement of your thermostat to make positive that when the ground machine arrives, the cold lead wires will obtain your thermostat location.

Floor mats the locations place

Determine whether or not or now not you can use elegant dimension off the shelf heating mats or if you want a custom-made design. Custom designs continuously cowl the floor area better; on the other hand many small rectangular or rectangular rooms can be protected pretty precise with today's mats.

Clean the floor and insert mats

Make sure the subfloor is clean and free of debris. With a modern subfloor you can additionally select to set up screws in the floor joists to cease squeaks and make certain that it's successfully fastened. Avoid the temptation to tear open the area and commence inserting in your mat barring first analyzing through the pointers and dry-fitting your electric powered heating system. Heated Tile Floor, install a tile backer board as an underlayment if placing in tile.

Lay the floor cover

Conduct insulation and resistance assessments of the electrical system, and then proceed with laying the performed floor covering. Thin-set mortar technique Depending on the product stable mat or mesh mat, you may also moreover have to study thin-set mortar first. For robust mats, exercise the mortar with a 1 / 3-by-1 / 3-in. rectangular notch trowel and then set the sturdy mat into the thin-set.

Mortar floor and dry

Use a rubber go with the flow to press the entire mat flooring to firmly create 100-percent contact between the heating mat, the mortar and the subfloor. Use a 1/4″ x 1/4″ trowel to lay the thin-set for the steady mats. Be sure to dry-fit the heating mats formerly than placing in with mortar. If the product is a mesh mat you can tightly closed the mesh mats to the sub-floor and then observes thin-set over the warmness cables the utilization of the flat aspect of the trowel.

Thin set and round up

Hot glue can moreover be used to tightly close the mesh to the floor. It would possibly additionally aid to butter the backs of the tiles with thin-set for greater adhesion to the mortar bed.

How To Reduce Electricity Bills

Beef up your insulation to keep the cold air in. Get a programmable thermostat to alter temperatures when you're away or sleeping. Maintain your air conditioner to keep it going for walks effectively and make sure a prolonged lifespan.

Use thick curtains or blinds to maintain the solar from coming in and heating your home. Use ceiling followers to help drift into the cool air and put a great deal much less stress on your AC.

Complete Guide To Repair A/C Leaking

Close observe coils

Look for Frozen Evaporator Coils, when you first are conscious a water leak coming from your A/C unit, you'll first want to take a look at the evaporator coil due to the reality that is the most widely wide-spread problem buildings experience. The evaporator coil seems like a coiled tube with fins. If you word ice on this component, shut down your A/C unit straight away formerly than in addition damage or an inaccurate computer ensues.

Flip off A/C units

Keep in idea that you ought to in no way try to chip away the ice on your own, as this can severely harm the fins on the evaporator coil. Even if this element is no longer usually frozen, you will although desire to flip off your A/C unit to maintain away from greater steeply-priced repairs and water harm mainly if the unit is located on the roof or in an attic or crawl residence the area dripping water can smash partitions and ceilings.

Locate the overflow drainage

Check for a Cracked or Overflowing Drain Pan, if your evaporator coil looks to be intact, the drain pan is placed perfect under your unit's evaporator coil, and its predominant attribute is to acquire condensation from your indoor A/C unit. If the drain pan is cracked, rusted, or overflowing, this functionality water is now not accurate draining outside, inflicting a build-up and leaks. Keep in wondering that you can rapidly restoration an overflowing drain pan the utilization of water sealant.

Clean and seal the linkage

Clean a Clogged Condensate Drain Line, over time algae, fungi, and exclusive particles can clog your A/C unit's condensate drain line. Pouring 6 oz vinegar into drain line few months clear out any build-up of algae, fungi, or specific matter.

Common Issues Of Air Condition

Issues of mini breakdown AC

When your air conditioning unit unexpectedly shuts down with no warning,

Solutions

First, check your thermostat batteries. An easy-to-forget household chore is automatically altering machine batteries, and thermostats are frequently forgotten about. If the batteries are dead, alternate them. Then, affirm that it is set to cooling and the temperature you desire. If all else fails, you may additionally moreover choose to reset the AC's circuit breaker.

Issues of Blowing Warm or Hot Air

There are few things larger uncomfortable than an air conditioner that dispenses lukewarm air. Before doing anything, verify that your thermostat settings weren't via accident switched.

Solutions

If your settings are proper the first thing you ought to do is alternate your air filter. Old, clogged air filters are frequently to blame for AC problems like frozen coils. If the filter isn't the problem; you have to be low on refrigerant.

Test this with the aid of way of feeling the giant of the two copper traces that go into the unit's condenser. If it is moist and bloodless to the contact your tiers are good. If it's lacking in one or every of these qualities, refrigerant is low. Refilling or repairing the refrigerant reservoir.

Issues of Leakage

Air conditioning gadgets are predicted to include a fine volume of fluid and condensate, alternatively excessive leaking is indicative of limitless one-of-a-kind troubles with numerous solutions.

Solutions

The condensate line can barring subject flip out to be clogged. Getting it unclogged is rudimentary fix; on the other hand make sure you have a look at the manufacturer's instructions.

Issus of high Temperature

The hassle ought to be an unbalanced air system. Depending on room size, air vent placement, windows, and even exterior temperature some areas of your home won't get as masses activity from the AC unit.

Solutions

Make sure massive domestic home windows are included with sun-blocking curtains or blinds. From here, you can take a seem at the insulation of each and every room or set up dampers to balance your computer and redistribute some of the air goes with the float to preserve each and every room equally cool.

Issue of excess sound

You want to be in a role to feel that your AC is working; alternatively you shouldn't have to hear it. If your AC sounds like it's struggling to keep you cool it perhaps is struggling. Failing air conditioners have a tendency to make extraordinary noises that are difficult for a novice to become aware of or diagnose.

Solutions

Screeching and squealing must mean there's an inaccurate belt.

Rattling sounds would possibly additionally be guide of problems with the fans, motor, or compressors. Beyond checking your thermostat, clearing particles from spherical you're out of doorways unit, and altering the filters.

How Thermostat works in supplement to A/C

The temperature sensor on a thermostat suggests when the heater or air conditioner have to be running or grew to become off. There may additionally prefer to be varying of thermally managed zones, each of which requires their very personal thermostat. The thermostat ought to be positioned someplace as some distance as doable from areas of centered temperature difference with the mean temperature of the intended space.

What is Heat Generator?

Heat generator is the key member of HVAC gadget factors when it comes to heating. What takes vicinity in these devices is the technological know-how of heat, for instance, by extraction of gas electrical energy interior a furnace, aka combustion chamber.

How fluid and electric gadget

Hot flue gases will then furnish heating for the air or each and every different fluid such as water that will later warmness the air getting into the conditioned environment. Electric warmness technological know-how ought to moreover be used to heat the conditioning air.

Although there may also be a vary of picks for heat generators, the most normal sorts are the furnaces, and therefore, it is crucial to mirror on consideration on combustion for beneficial useful resource manipulates and pollutant emission for environmental concerns involving these HVAC computing device components.

What is The Heat Exchanger?

Heat exchangers are one of HVAC desktop factors that gain the warmness generated in the heat technological know-how unit and change it to each and every different fluid.

How heat exchangers discharge it duties

Some manipulate unit will set off the furnace or electric powered heating elements when wished to alter the air temperature passing through the warmness exchanger.

In many applications, the warmness is transferred besides lengthen to the cool air in order to grant heating for the supposed space. In this case, some device blows the air via heat flue gasoline tubes or electric powered heating elements, transferring the strength to the air by using way of warmness absorption.

What is Blower and it performance?

The air is compelled via one of the HVAC machine components, acknowledged as the blower, via the heat exchanger into the air ducting that would take the warmth air to the area it is intended. The blower is pushed by way of ability of an electric powered motor thru a shaft. The float of the air ought to regulate by way of capability of enhancing the motor speed. Such motors want to be of the variable tempo type.

Basic role of Condenser Coil in HVAC

The Condenser Coil, One of the critical HVAC system elements is the compressor or condenser coil which is typically located outside. The warmness refrigerant fuel is taken to the compressor to dissipate warmness to the outdoor surroundings and flip into its liquid form. This liquid refrigerant is then taken to the evaporator coil via copper or aluminum tubes. A fan will amplify the extent of air flown passed the coils and make bigger the condensation process.

What is Evaporator Coil?

The evaporator coil is one of HVAC gadget factors located indoors that receives the condensed refrigerant liquid from the compressor. The liquid refrigerant is atomized through way of spraying nozzles that will increase the fee of refrigerant evaporation when it comes to contact with the room's warmness air. There are followers that make the room's warmness air flow by means of the return ducts onto the evaporator.

How *Evaporator Coil* discharge it functions

The warmth air rejects heat to the atomized refrigerant and cool down, after which it is redistributed lower back to the rooms by using the ducting. As the air passes over the cold evaporator coil, its moisture stage would be lowered due to condensation of the moist air on the coil.

Air Ducts and Vents as a component unit of HVAC

The air is transferred through ducts to obtain specific HVAC system components. Good ducting is essential to have immoderate exquisite air delivered to the zone. Duct leakage ought to quit end result in noise when the system is working. In addition, when the air ducting is now not in proper shape, scent and more moisture ought to fill the air.